CASEY TREES
SPECIES GUIDE

EASY-TO-USE REFERENCE FOR
THE MOST COMMONLY FOUND TREE SPECIES
IN WASHINGTON, DC

Casey Trees®
WASHINGTON DC

CONTENTS

ABOUT THIS GUIDE

The mission of Casey Trees is to restore, enhance, and protect the tree canopy of our nation's capital. Attaining that mission depends on the support of an engaged citizenry. We hope this guide helps to deepen your involvement with, and enrich your understanding of, the fascinating world of trees in and around Washington, DC.

This guide showcases the 56 trees most commonly found in DC's parks. It was written primarily for those new to tree identification and is informed by years of practical experience gained by Casey Trees staff teaching tree identification to volunteers. The document avoids highly technical terms, contains helpful illustrations, identifies trees that are commonly mismatched, and provides interesting anecdotes.

DC's trees — and trees in urban areas generally — provide many benefits: reducing summer heat, improving air quality, slowing damaging flood waters, or simply making neighborhoods more beautiful and inviting. Visit *caseytrees.org* to learn more about how you can help Casey Trees make a difference. Thank you for your support!

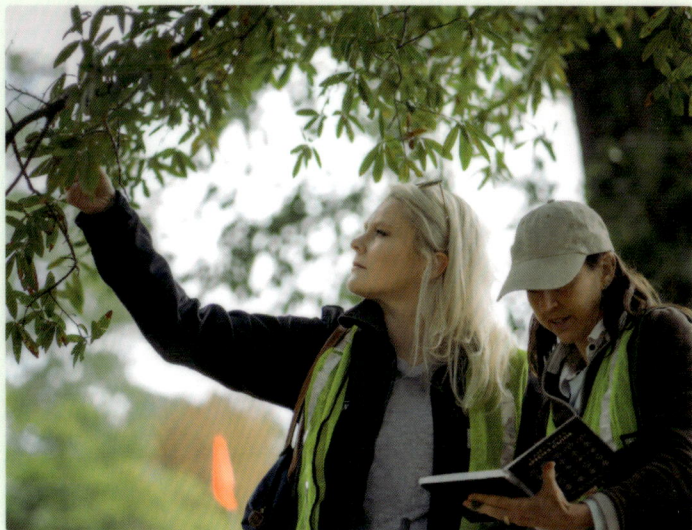

USING THIS GUIDE

This guide is best used when leaves and/or fruits/flowers are available to see. First, examine the leaves of the tree and compare in the beginning leaf guide. If you have fruits, compare those in the fruit guide. Then you will be directed to a particular species profile.

Key characteristics for tree ID are overall form, leaf shape, leaf attachment, leaf arrangement, flowers/fruit, and bark.

LEAF STRUCTURES

LEAF MARGIN (EDGE)

ENTIRE / SMOOTH

No teeth, serrations, or lobes. Smooth all the way around.

SERRATED / TOOTHED

Continuous, sharp, forward-facing edges.

LOBED

Pointed or round ins and outs that don't reach the center vein.

LEAF ARRANGEMENT

ALTERNATE

Leaves alternate from side to side, with each leaf coming off at a different point on the stem.

OPPOSITE

Leaves are arranged across from one another on the stem.

LEAF VENATION

PINNATE

Smaller, lateral veins extend out of one central vein.

PALMATE

Major veins radiate out of one central location.

LEAF ATTACHMENT

SIMPLE

One uninterrupted leaf with a bud at its base.

COMPOUND

One interrupted leaf with a bud at its base and leaflets extending off the central vein.

FEATURED SPECIES

NEEDLE / SCALE-LIKE / EVERGREEN
Arborvitae · *Thuja occidentalis*
Eastern Red Cedar · *Juniperus virginiana*

NEEDLE / SIMPLE OR IN CLUSTERS / EVERGREEN
Japanese Cryptomeria · *Cryptomeria japonica*
Eastern White Pine · *Pinus strobus*
Virginia Pine · *Pinus virginiana*
Loblolly Pine · *Pinus taeda*
Spruce · *Picea spp.*

NEEDLE / SIMPLE / DECIDUOUS
Bald Cypress · *Taxodium distichum*

FAN LEAF / SIMPLE / DECIDUOUS
Ginkgo · *Ginkgo biloba*

LEAF / SIMPLE / EVERGREEN
Holly · *Ilex opaca*
Southern Magnolia · *Magnolia grandiflora*
Sweetbay Magnolia · *Magnolia virginiana*
Saucer Magnolia · *Magnolia x soulangeana*

LEAF / SIMPLE / ALTERNATE / LOBED / SMOOTH
Scarlet Oak · *Quercus coccinea*
Black Oak · *Quercus velutina*
Northern Red Oak · *Quercus rubra*

Pin Oak • *Quercus palustris*

Shumard Oak • *Quercus shumardii*

Eastern White Oak • *Quercus alba*

Chestnut Oak • *Quercus montana*

Bur Oak • *Quercus macrocarpa*

Swamp White Oak • *Quercus bicolor*

Overcup Oak • *Quercus lyrate*

Tulip Tree • *Liriodendron tulipifera*

LEAF / SIMPLE / ALTERNATE / SMOOTH

Crape Myrtle • *Lagerstroemia spp.*

Persimmon • *Diospyros virginiana*

Blackgum or Tupelo • *Nyssa sylvatica*

Pawpaw • *Asimina triloba*

Willow Oak • *Quercus phellos*

Redbud • *Cercis canadensis*

Sourwood • *Oxydendrum arboreum*

Sassafras • *Sassafras albidum*

LEAF / SIMPLE / ALTERNATE / LOBED / SERRATED

Mulberry • *Morus spp.*

Sweetgum • *Liquidambar styraciflua*

Sycamore and London Planetree • *Platanus spp.*

LEAF / SIMPLE / ALTERNATE / SERRATED

Crab Apple • *Malus spp.*

Weeping Willow • *Salix babylonica*

River Birch • *Betula nigra*

Hophornbeam • *Ostrya virginiana*

Hornbeam · *Carpinus caroliniana*
Yoshino Cherry · *Prunus x yedoensis*
Kwanzan Cherry · *Prunus serrulata*
Weeping Cherry · *Prunus pendula*
Elm · *Ulmus americana*
American Basswood · *Tilia americana*
Littleleaf Linden · *Tilia cordata*
Silver Linden · *Tilia tomentosa*
Chinese Elm · *Ulmus parvifolia*
Serviceberry · *Amelanchier spp.*
Callery Pear · *Pyrus calleryana*
Beech · *Fagus grandifolia*
Sawtooth Oak · *Quercus acutissima*
Hackberry · *Celtis occidentalis*
Japanese Zelkova · *Zelkova serrata*

LEAF / SIMPLE / OPPOSITE / SMOOTH

Catalpa · *Catalpa speciosa*
Flowering Dogwood · *Cornus florida*
Kousa Dogwood · *Cornus kousa*
Alternate-Leaf Dogwood · *Cornus alternifolia*
Fringe Tree · *Chionanthus virginicus*

LEAF / SIMPLE / OPPOSITE / SERRATED

Red Maple · *Acer rubrum*
Sugar Maple · *Acer saccharum*
Silver Maple · *Acer saccharinum*

LEAF / COMPOUND / PINNATE / ALTERNATE / SMOOTH

Yellowwood · *Cladrastis kentukea*

Black Locust · *Robinia pseudoacacia*

Japanese Pagoda · *Sophora japonica*

Chinese Pistache · *Pistacia chinensis*

LEAF / COMPOUND / ALTERNATE / SERRATED

Black Walnut · *Juglans nigra*

LEAF / COMPOUND / OPPOSITE / SERRATED

Boxelder · *Acer negundo*

Buckeye · *Aesculus glabra*

Black Ash · *Fraxinus nigra*

Carolina Ash · *Fraxinus caroliniana*

Green Ash · *Fraxinus pennsylvanica*

White Ash · *Fraxinus americana*

LEAF / COMPOUND / BI-PINNATE / ALTERNATE / SMOOTH

Honey Locust · *Gleditsia triacanthos*

Kentucky Coffeetree · *Gymnocladus dioicus*

LEAF / COMPOUND / BI-, TRI-PINNATE / ALTERNATE / SERRATED

Golden Raintree · *Koelreuteria paniculata*

LEAF GUIDE

NEEDLE / SCALE-LIKE / EVERGREEN

Arborvitae • page 28

Eastern Red Cedar • page 29

NEEDLE / SIMPLE OR IN CLUSTERS / EVERGREEN

Japanese Cryptomeria • page 30

Virginia Pine • page 32

Loblolly Pine • page 32

Eastern White Pine • page 31

Spruce • page 33

NEEDLE / SIMPLE / DECIDUOUS

Bald Cypress • *page 34*

FAN LEAF / SIMPLE / DECIDUOUS

Ginkgo • *page 35*

LEAF / SIMPLE / EVERGREEN

Holly • *page 36*

Magnolias • *pages 37-39*

LEAF / SIMPLE / ALTERNATE / LOBED / SMOOTH

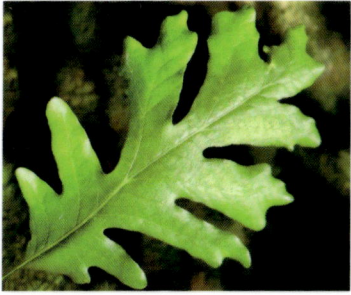

Some Oaks • *pages 40-45*

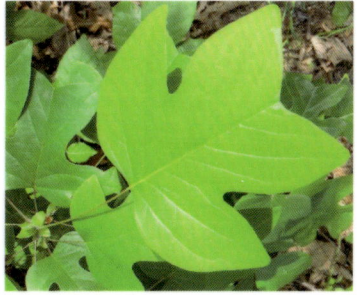

Tulip Tree • *page 46*

LEAF / SIMPLE / ALTERNATE / SMOOTH

Crape Myrtle • *page 48*

Persimmon • *page 58*

Blackgum/Tupelo • *page 49*

Pawpaw • *page 47*

Willow Oak • *page 43*

Redbud • *page 50*

Sourwood • *page 52*

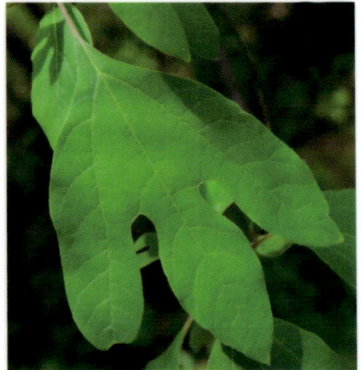

Sassafras • *page 51*

LEAF / SIMPLE / ALTERNATE / LOBED / SERRATED

Mulberry • *page 57*

Sweetgum • *page 59*

Sycamore and London Planetree • *page 60*

LEAF / SIMPLE / ALTERNATE / SERRATED

Crabapple • *page 61*

Weeping Willow • *page 62*

River Birch • *page 65*

Hophornbeam • *page 67*

LEAF / SIMPLE / ALTERNATE / SERRATED (CONT'D)

Hornbeam • *page 66*

Cherry • *page 68*

Elm • *page 70*

Linden • *page 71*

Chinese Elm • *page 76*

Serviceberry • *page 73*

Callery Pear • *page 74*

Beech • *page 63*

Sawtooth Oak • *page 45*

Hackberry • *page 64*

Japanese Zelkova • *page 75*

LEAF / SIMPLE / OPPOSITE / SMOOTH

Catalpa · *page 53*

Dogwood · *page 54*

Fringe Tree · *page 56*

LEAF / SIMPLE / OPPOSITE / SERRATED

Red Maple · *page 78*

Sugar Maple · *page 79*

Silver Maple · *page 80*

LEAF / COMPOUND / PINNATE / ALTERNATE / SMOOTH

Yellowwood · *page 77*

Black Locust · *page 81*

LEAF / COMPOUND / PINNATE / ALTERNATE / SMOOTH (CONT'D)

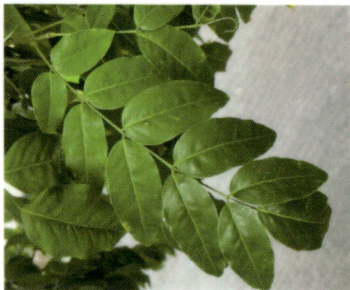

Japanese Pagoda • *page 82*

Chinese Pistache • *page 83*

LEAF / COMPOUND / ALTERNATE / SERRATED

Black Walnut • *page 84*

LEAF / COMPOUND / OPPOSITE / SERRRATED

Buckeye • *page 88*

Ash • *page 86*

Boxelder • *page 85*

LEAF / COMPOUND / BI-PINNATE / ALTERNATE / SMOOTH

Honey Locust • *page 89*

Kentucky Coffeetree • *page 91*

LEAF / COMPOUND / BI-,TRI-PINNATE / ALTERNATE / SERRATED

Golden Raintree • *page 90*

FRUIT GUIDE

Arborvitae • *page 28*

Japanese Cryptomeria • *page 30*

Eastern White Pine • *page 31*

Virginia Pine • *page 32*

Loblolly Pine • *page 32*

Spruce • *page 33*

Bald Cypress, • *page 34*

Magnolias • *pages 37-39*

BERRY-LIKE

Eastern Red Cedar • *page 29*

Holly • *page 36*

Fringe Tree • *page 56*

Blackgum • *page 49*

Hackberry • *page 64*

Dogwoods • *page 54*

Mulberry • *page 57*

Serviceberry • *page 73*

Japanese Zelkova • *page 75*

FLESHY CIRCULAR FRUIT

Crabapple • *page 61*

Cherry • *page 68*

Gingko • *page 35*

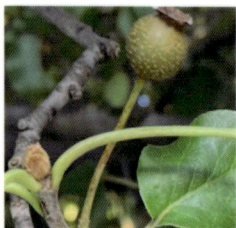
Callery Pear • *page 74*

Pawpaw • *page 47*

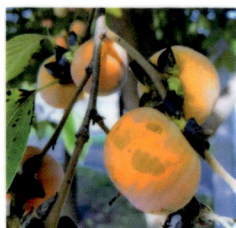
Persimmon • *page 58*

ACORN OR NUT

Oaks • *pages 40-45*

Beech • *page 63*

Buckeye • *page 88*

Chinese Pistache • *page 83*

Black Walnut • *page 84*

SEED GATHERING OR CLUSTER

Crape Myrtle • *page 48*

Linden • *pages 71-72*

Sweetgum • *page 59*

Sycamore and London Planetree • *page 60*

River Birch • *page 65*

Hophornbeam • *page 67*

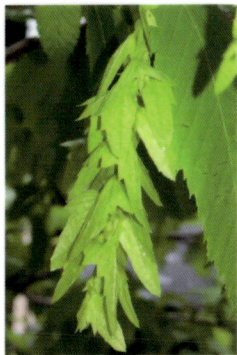

Hornbeam • *page 66*

SAMARA (WINGED FRUIT)

Elm • *page 70*

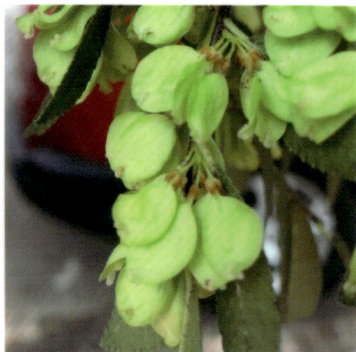

Chinese Elm • *page 76*

Maples • *pages 78-80*

Boxelder • *page 85*

Ash • *page 86*

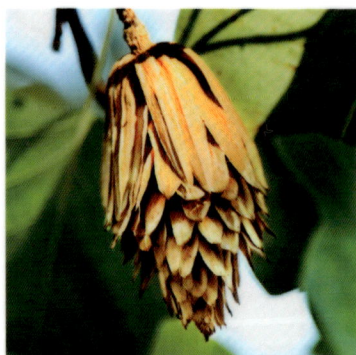

Tuliptree • *page 46*

SEED POD OR BEAN-LIKE

Redbud • *page 50*

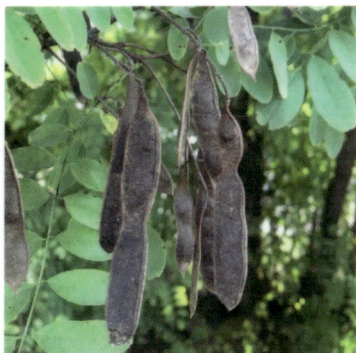

Black Locust • *page 81*

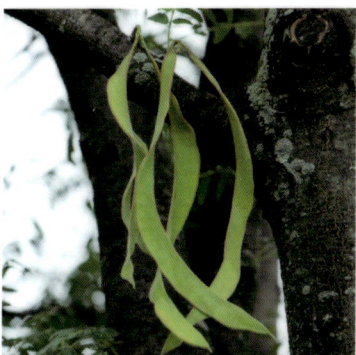

Honey Locust • *page 89*

Japanese Pagoda • *page 82*

Yellowwood • *page 77*

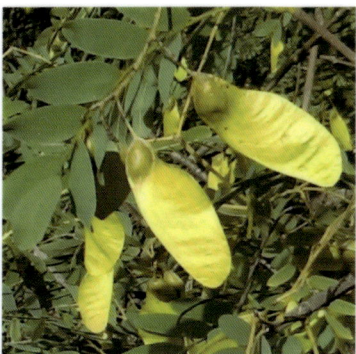

Kentucky Coffeetree • *page 91*

ARBORVITAE · *Thuja occidentalis*

LEAF ARRANGEMENT	LEAF PERSISTENCE	TREE SHAPE
Simple	Evergreen	Pyramidal, Oval

NEEDLE	BARK
Flattened, scale-like, triangular ⅜" needles, overlapping and 2-ranked, with short points; glossy, emerald green	Fibrous, stringy, brown-red to gray

CONE	NOTES
Round with woody scales, forming at branch tips, purple-brown and waxy to gray-brown upon maturity	Dense crown, 15–20' tall, popular landscaped tree for hedges due to columnar form

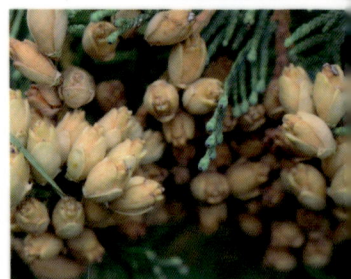

EASTERN RED CEDAR · *Juniperus virginiana*

LEAF ARRANGEMENT	LEAF PERSISTENCE	TREE SHAPE
Simple	Evergreen	Conical, Narrow Pyramid

NEEDLE	BARK
Very small, rounded, scale-like 1/16" needles, dark blue-green turning lighter green upon maturity	Shaggy, reddish-brown, soft and silvery where exposed

CONE	NOTES
Berry-like, powdery cones, pale green turning to a dusty blue, pea-sized and mature in one spring–fall growing season	Native coniferous evergreen often used as a natural screen in landscaping

MAY BE CONFUSED WITH

CYPRESS, JUNIPER, FALSE-CYPRESS, ARBORVITAE, INCENSE CEDAR · Distinguished from these species by fleshy, powdery blue berry

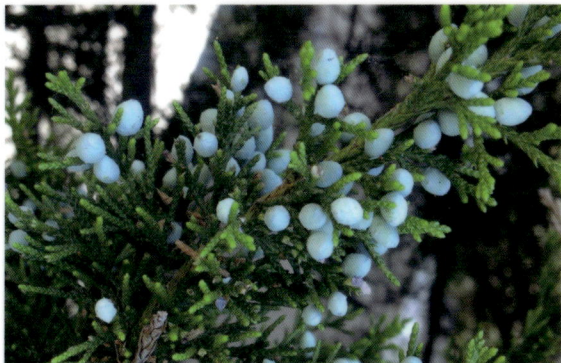

JAPANESE CRYPTOMERIA · *Cryptomeria japonica*

LEAF ARRANGEMENT	LEAF PERSISTENCE	TREE SHAPE
Simple	Evergreen	Pyramidal, Oval

NEEDLE	BARK
Thin, pointed, ½–¾", prickly leaves, spirally arranged, curving toward twig, green but may turn bronze or brown in winter	Peeling in vertical stripes, stout, red-brown

CONE	NOTES
Separate male cones appearing at leaf axils as light brown, spherical fruiting cones and female cones appearing as round, spiky, ¾" cones with multiple pointed scales, green to reddish-brown	Leaf arrangements hang down loosely giving uniform branch appearance. Tree 50–80' tall, 20–30' wide.

EASTERN WHITE PINE · *Pinus strobus*

LEAF ARRANGEMENT	LEAF PERSISTENCE	TREE SHAPE
Simple	Evergreen	Irregular, Oval

NEEDLE	BARK
Thin, straight, 4" needles, blue-green, 5 to each bunch	Light gray bark becomes darker, thicker with long ridges and dark furrows with a hint of purple upon maturity

CONE	NOTES
Young male cones are yellowish ½" pellets that bunch near branch tips and young female cones are 1-1 ½" long, light green with hints of red at the branch tips; mature cones are long, slender cones, 5 ½" with sappy scales, mature in late summer	Triangular outward gathering of needles at branch tips

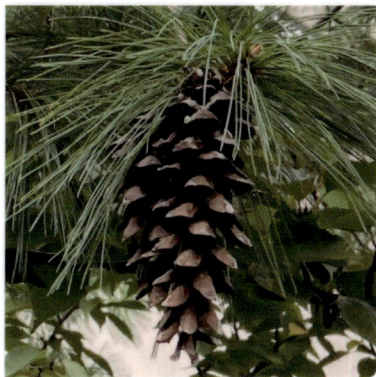

VIRGINIA PINE •
Pinus virginiana

LEAF ARRANGEMENT	LEAF PERSISTENCE
Simple	Evergreen

TREE SHAPE	NEEDLE
Irregular, Oval	Small, twisting, 2"–3" needles, 2 per bundle

BARK
Rough, irregular scaled, gray below and redder above branching

CONE
Small, mildly prickly cones, 2–3", open upon maturity

NOTES
Flexible branching, irregular branching often with no central leader, between 20–40' tall

LOBLOLLY PINE •
Pinus taeda

LEAF ARRANGEMENT	LEAF PERSISTENCE
Simple	Evergreen

TREE SHAPE	NEEDLE
Irregular, Oval	Thin 7" needles, average of 3 per bundle, bright green

BARK
Red-brown, very furrowed and rough with large rectangular scales

CONE
Oval, 3 – 6" with prickly scales

NOTES
Extremely fast growing, losing its lower limbs as it grows, between 60 – 90' tall

SPRUCE · *Picea spp.*

LEAF ARRANGEMENT	LEAF PERSISTENCE	TREE SHAPE
Simple	Evergreen	Conical, Pyramidal

NEEDLE	BARK
Sharp, stiff, ½–1 ¼" needles, expelling straight from twig in spiraling fashion, silvery blue-green to dark green	Gray to pale red-brown, with thin scales

CONE	NOTES
Mature cones are cylindrical with wavy, flexible scales 2–4", and hanging down; young cones turning from pink to light brown, sitting upright	Medium to large trees with branches drooping, can reach 80–100' tall

MAY BE CONFUSED WITH

PINE, FIR, HEMLOCK or **DOUGLAS-FIR** · Distinguished by wavy, scaled cones

BALD CYPRESS · *Taxodium distichum*

LEAF ARRANGEMENT	LEAF PERSISTENCE	TREE SHAPE
Simple	Deciduous	Pyramidal

NEEDLE	BARK
Flat, soft ¾" needles, generally 2 rowed, yellow-green; alternate on branch	Reddish-brown turning gray, vertically shredding upon maturity; identifiable by their knees (knobby masses above the ground)

FRUIT	NOTES
Mature spherical cones, 1", composed of shield-shaped scales, woody, green maturing to brown; male cones are rust-colored grouped catkins, 7", appearing before leaves start to emerge in spring	Common in wet, swampy habitat, usually between 50–60' tall, but can reach 150' in some areas; sparse, feathery branching

MAY BE CONFUSED WITH

DAWN REDWOOD · Larger, 1" wide needles in dense flat sprays; small, feathery, four-sided, small, cylindrical cones? It could be a Dawn Redwood!

ENGLISH YEW · Large shrub or small tree; bright, fleshy, red, berry-like fruit covering exposed seed? It could be an English Yew!

EASTERN HEMLOCK · Flat, ½" long, single needles that taper to a dull point, primarily two-ranked; cone ¾" with rounded scales? It could be an Eastern Hemlock!

GINKGO · *Ginkgo biloba*

LEAF ARRANGEMENT	LEAF PERSISTENCE	TREE SHAPE
Simple	Deciduous	Irregular, Oval Columnar

LEAF	BARK
Fan-shaped, 3" leaves, veins diverging from base	Light gray-brown with irregular flat ridges, eventually breaking into deeper, corky furrows

FRUIT	FLOWER
Circular, fleshy 1" fruit on long stalks, large pit, strong unpleasant odor upon dropping, maturing in fall following the first frost	Slim, inconspicuous (no petals) cylindrical flower (catkin-like) blooming in early spring

NOTES

Fossils show evidence of existence over 200 million years ago; 40–70' tall

HOLLY · *Ilex opaca*

LEAF ARRANGEMENT	LEAF PERSISTENCE	TREE SHAPE
Simple, Alternate	Evergreen	Dense Pyramidal, Columnar

LEAF

Stiff, curled, 3 ½" glossy green leaves, wide at base with sharp, blood-thirsty spikes emerging from edges, underleaf paler

BARK

Smooth, gray, resembling elephant skin

FRUIT

Small, ½" berries, arranged in small clusters, usually red or yellow (only present on female species)

FLOWER

Tiny, greenish-white, and appear on new growth (May–June)

NOTES

Dense, shrubby, small evergreen tree with short trunk and branches to ground, up to 40' tall

SOUTHERN MAGNOLIA · *Magnolia grandiflora*

LEAF ARRANGEMENT	LEAF PERSISTENCE	TREE SHAPE
Simple, Alternate	Evergreen	Irregular, Broadly Round

LEAF	FRUIT
Thick, rigid, dark green, 5–8" leaves, underleaf pale with rusty brown hairs	Oval green to red to brown, 3 ½" hairy cone

FLOWER	NOTES
Cup shaped, creamy white, citrus fragrance, 9–12 thick petals	Distinguished by underleaf and establishment as larger, single-trunked tree

SWEETBAY MAGNOLIA • *Magnolia virginiana*

LEAF ARRANGEMENT	LEAF PERSISTENCE	TREE SHAPE
Simple, Alternate	Evergreen	Irregular, Broadly Round

LEAF	FRUIT
Paler, shiny, rounded base, bluntly v-tipped, 3–5" leaves, dark green, whitish below	Pink to reddish-brown, 2" cone, expelling red seeds

FLOWER	NOTES
Smaller, creamy white, 9–12 petals, 2–3" flower, citrus fragrance	Long buds, silvery-gray, curling slightly; distinguished from other species by narrow leaf and smaller fruit

SAUCER MAGNOLIA · *Magnolia x soulangeana*

LEAF ARRANGEMENT	LEAF PERSISTENCE	TREE SHAPE
Simple, Alternate	Evergreen	Irregular, Oval

LEAF	FRUIT
Wavy-edged, mildly stiff, short-pointed 5" leaves, fine hairs below	2–3" cone, expelling red seeds

FLOWER	NOTES
Showy, 4–8", light pink to purple petals	Fuzzy buds, small, multi-stemmed with narrow crown; distinguished from Star Magnolia by smaller, many petaled (up to 18) white flowers

RED OAK GROUP · *Quercus spp.*

LEAF ARRANGEMENT	LEAF PERSISTENCE	TREE SHAPE
Simple, Alternate	Deciduous	Broadly Round

LEAF	TWIGS
Pointed lobes, some veins extend beyond margins with bristle-tips	Larger buds with a pointed tip in comparison to white oaks

BARK	FRUIT
Furrowed, darker, and more concrete than white oaks	Cap has flat scales and is hairy on the inside. Mature in the fall following a 15–16 month growing cycle

FLOWER

Separate male and female flowers on the same tree; male flowers long, catkins at the end of the twig, and female flowers small and sessile on branch where acorns form

WHITE OAK GROUP · *Quercus spp.*

LEAF ARRANGEMENT	LEAF PERSISTENCE	TREE SHAPE
Simple, Alternate	Deciduous	Broadly Round

LEAF		TWIGS
Rounded lobes, lacking bristle-tips		Smaller and more rounded buds than red oaks

BARK	FRUIT
Scaly, lighter, and peeling easily	Cap has raised scales and is not hairy on the inside. Mature in the fall after a 4–6 month growing cycle

FLOWER

Separate male and female flowers on the same tree; male flowers long, catkins at the end of the twig, and female flowers small and sessile on branch where acorns form

SCARLET OAK ★
Quercus coccinea

LEAF
Bristle dipped, prominent middle lobe, 6" leaves, 7 deep lobes

FRUIT
Shiny acorns, ½–1", deeply capped, (⅓ - ½ of acorn), might display rings or cracks at cap point

NOTES
Differs from Pin Oak with deeper acorn cap, pale orange-brown tufts at central vein joining, and rough bark

★ OFFICIAL TREE OF WASHINGTON, DC

BLACK OAK ·
Quercus velutina

LEAF
Variable shape, sun leaves with deep sinuses versus shade leaves with shallow sinuses, 7 ½" leaves, 5–9 bristle-tipped lobes, underleaf slight orange tint

FRUIT
Acorns ¾", bowl-shaped cap, scales loose especially on edge of cap, slightly fuzzy

NOTES
Larger leaf with broader sinuses and darker bark than Northern Red Oak

NORTHERN RED OAK ·
Quercus rubra

LEAF
Tapered, pointed with shallow lobes, 7" leaves, 7–11 lobes; leaf stalks frequently red, underside smooth, gently drooping from twigs

NOTES
Acorns 1", caps thin and tight; in clusters of 2-5 or single

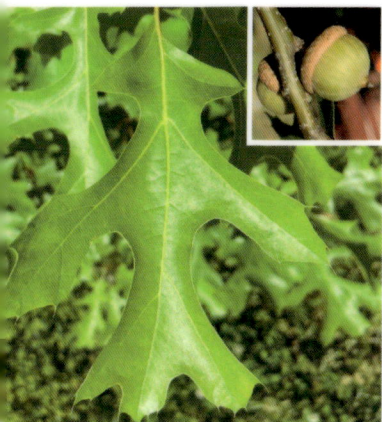

PIN OAK · *Quercus palustris*

LEAF
Bristle-tipped lobes, with sinuses reaching almost to midrib, 5 ½" leaves, 5–9 deep, major lobe sinuses; U-shaped

FRUIT
Acorns small, ½", rounded, thin, flat cap

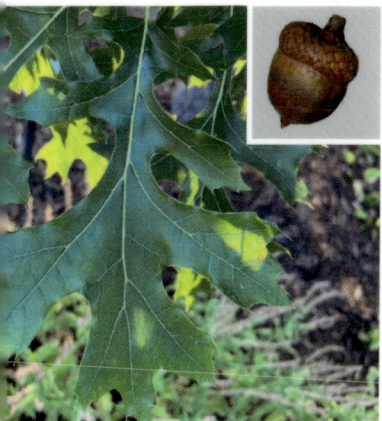

SHUMARD OAK · *Quercus shumardii*

LEAF
Squarish, bristle-tipped 5 ½" leaves, 5–9 deep lobes with sinuses reaching halfway to midrib

FRUIT
Acorns ¼" long, rounded, with flattened, bowl-like caps, short stalk

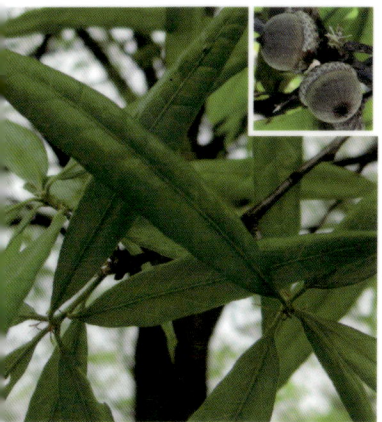

WILLOW OAK · *Quercus phellos*

LEAF
Elongated, narrow, spear-shaped, 4" leaves with short stalks, entire margins, and bristle tip

FRUIT
Very small ¼ – ½" acorns, yellow-green, nearly circular; caps are saucer-like with thin, shallow scales, covering only ¼ of the nut

EASTERN WHITE OAK •
Quercus alba

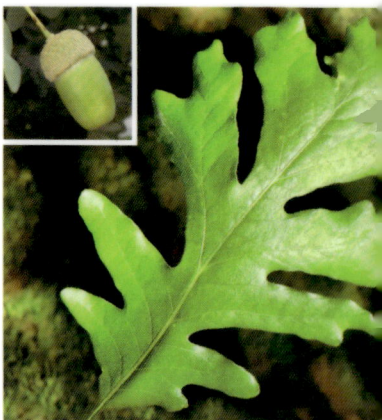

LEAF

Variable shape, deeply tapered base and medium-large rounded lobes, 6 ½. " leaves, matte and more gray-blue than other oaks

FRUIT

Acorns ¾" with thin caps and short stalks

NOTES

Most common of White Oaks in DC

CHESTNUT OAK •
Quercus montana

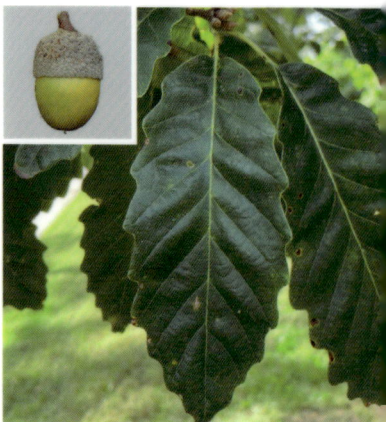

LEAF

Broad, 6" leaves, 7–17 pairs of rounded teeth, underleaf pale green

FRUIT

Acorns 1 ⅛" long, warty, teacup shaped, fairly deep caps, often separating when mature

BUR OAK •
Quercus macrocarpa

LEAF

Variable shapes, resembling fiddle, blunt tipped, 5 ½" leaves, middle of leaf deeply lobed, reaching midrib

FRUIT

Acorns large, 2", deeply fringed and shaggy cap

SWAMP WHITE OAK ·
Quercus bicolor

LEAF

Broad, non-uniform teeth, blunted shallow lobes, 6 ½" leaves, underleaf is whitish

FRUIT

Acorns 1" with long, slim stalks, and large tufted bowl-shaped caps

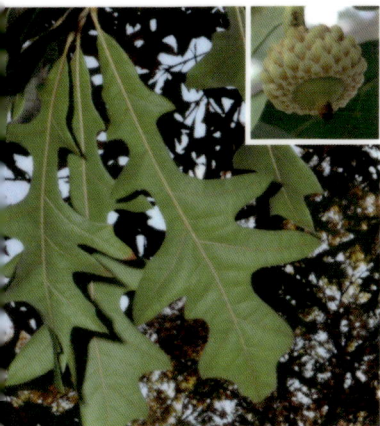

OVERCUP OAK ·
Quercus lyrata

LEAF

Variable shape, often with 3 equal, squarish lobes, broad sinuses at mid-leaf, 5 ½" leaves with narrow base, underleaf sometimes hairy

FRUIT

Large 1" acorn with scaly caps enclosing almost entire nut, acorns float on water

SAWTOOTH OAK ·
Quercus acutissima

LEAF

Skinny, elongated, 5 ½" leaves, with long bristle tips

FRUIT

Acorns 1", nearly round, cap covers about ⅔ of nut, cap is thick and has long, curved scales resembling hairs

TULIP TREE · *Liriodendron tulipifera*

LEAF ARRANGEMENT	LEAF PERSISTENCE	TREE SHAPE
Simple, Alternate	Deciduous	Narrow Oval

LEAF
Oddly 4-lobed, tulip or duck-feet-shaped, 5" leaves, long stalked, bright to dark green

BARK
Light gray with green tint, creases of past branches, developing lightened diamond-shaped flat ridges upon maturity

FRUIT
Upright woody 2" cones, releasing 1 ½" samaras

FLOWER
Showy, neon, 1 ¾" tulip-like flower with 6 yellow, green, and orange petals appearing in early summer to late spring

NOTES
Twigs wrap upwards with long buds, expel a sweet and spicy fragrance when crushed, and leaves randomly orient along branches; 50–60' tall but can reach up to 100'+

PAWPAW · *Asimina triloba*

LEAF ARRANGEMENT	LEAF PERSISTENCE	TREE SHAPE
Simple, Alternate	Deciduous	Small, Sparse

LEAF	BARK
Elongated, short-pointed, narrow base, 9" leaves with parallel veins, fragrance of green pepper when crushed	Thin, smooth, gray-brown with raised warty lenticels and gray patches

FRUIT	FLOWER
Unique, edible, 4", lumpy, green fruit, resembling a cross between a banana and a mango, soft texture, ripe when dropped in autumn	Dark purple-brown, 6-petal flower, resembles a bell, appearing in spring before leaves

CRAPE MYRTLE • *Lagerstroemia spp.*

LEAF ARRANGEMENT	LEAF PERSISTENCE	TREE SHAPE
Simple, Alternate	Deciduous	Vase

LEAF	BARK
Oval, 2 ½", leafstalk short, dark green above, paler below	Multi-stemmed, smooth, gray-brown, camo-like bark, peeling to expose reddish-brown and green

FRUIT	FLOWERS
Round fruit, ½", drying upright into 6 segmented, brown capsules, persisting through winter	Attractive, showy white, pink, red, or purple (depending on cultivar) flower clusters (10" long); appear in late summer

MAY BE CONFUSED WITH

BLACKGUM • Leaves with fine tips appear whorled around the branch, some leaves turning red; gray-brown bark; and blue-black berries? It's a Blackgum! (P. 49)

MAGNOLIA • Larger, glossy leaves, smooth gray bark; large cone like clustered fruit with red seeds, white tea-cup flowers? It's a Magnolia! (P. 37-39)

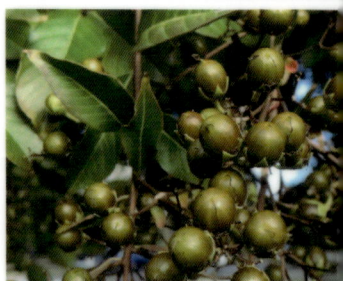

BLACKGUM or TUPELO · *Nyssa sylvatica*

LEAF ARRANGEMENT	LEAF PERSISTENCE	TREE SHAPE
Simple, Alternate	Deciduous	Oval

LEAF

Eye-shaped, 4" leaves, smooth margin generally, sometimes with rare serration near ends; appearing whorled along branches; glossy, bright to dark green and paler underneath; new leaves cluster vertically at branch ends; often display red leaves before fall

BARK

Smooth, gray, vertically furrowed, becoming chunky upon maturity

FRUIT

Dark blue or black berry; 1–5 (usually 3), on long reddish stalks

FLOWER

Small yellow-green male flowers in dense clusters and females in loose, open clusters; blooms in spring, with leaf out

MAY BE CONFUSED WITH

DOGWOODS · Veins curving to follow leaf edges; bark breaking into square blocks; and red fruit or blue-black berries? It's a Dogwood! (P. 54)

YELLOWWOOD · Lemon-shaped leaves, 3", 5–11 leaflets per leaf; and a flat green or brown pea-pod? It's a Yellowwood! (P.77)

MAGNOLIA · Larger, glossy leaves, smooth gray bark; large cone like cluster fruit with red seeds; showy, white-petaled flower? It's a Magnolia! (P. 37-39)

REDBUD · *Cercis canadensis*

LEAF ARRANGEMENT	LEAF PERSISTENCE	TREE SHAPE
Simple, Alternate	Deciduous	Broad Branching

LEAF
Round to heart-shape, thin, papery, 3–5" leaves, hanging down, smooth margin, green above, paler below

BARK
Smooth, gray-brown with orange furrows, scaly ridges, becoming darker gray upon maturity, often with maroon-orange patched cracks

FRUIT
Flattened fruit pod, 2–4", green, bean-like, turning brown

FLOWER
Small, purple-pink peak in clusters along branch, emerge in early spring before leaves, making tree appear pink

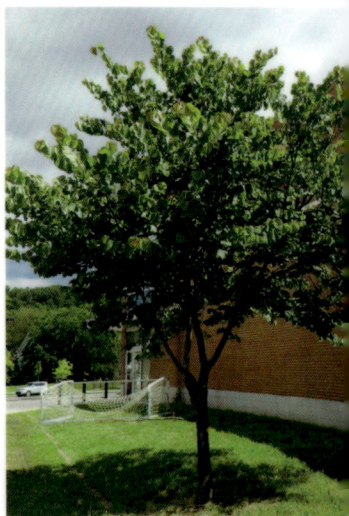

SASSAFRAS · *Sassafras albidum*

LEAF ARRANGEMENT	LEAF PERSISTENCE	TREE SHAPE
Simple, Alternate	Deciduous	Small, Irregular, Branching

LEAF
Exhibits 3 different leaf shapes; oval, double-lobed "mitten shape," tri-lobed "dinosaur foot shape"

BARK
Mahogany-brown in color on mature trees and rough, deeply furrowed

FRUIT
Oval, dark blue berries ~¼" on red stalk

FLOWER
Small, bright, greenish-yellow female flowers, clustered, appear before leaves in spring

NOTES
Height between 30–60' tall; single trunk tree or a dense, bushy thicket; aromatic bark and leaves

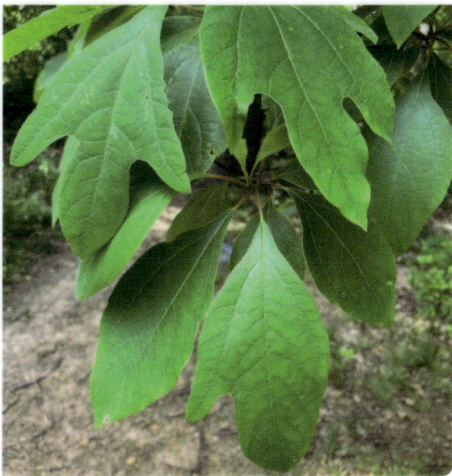

SOURWOOD · *Oxydendrum arboreum*

LEAF ARRANGEMENT	LEAF PERSISTENCE	TREE SHAPE
Simple, Alternate	Deciduous	Oval, Pyramidal

LEAF	BARK
Glossy, finely toothed, dark green leaves turning orange to red in fall, narrow, leaf 3–8"	Furrowed bark with broad ridges

FRUIT	FLOWER	NOTES
Small, beige, 5-angled capsules	Fragrant, white, small, urn-shaped flowers emerging from branch tips	Grows to be 50–60' tall

CATALPA · *Catalpa speciosa*

LEAF ARRANGEMENT	LEAF PERSISTENCE	TREE SHAPE
Simple, Opposite	Deciduous	Irregular

LEAF	BARK
Large, flexible, soft, pinnately veined, 8" leaves with long stalk, attached in whorls; light green to green with a hairy underside	Twisting gray to red-brown with shallow ridges

FRUIT	FLOWER
Long, slender, bean-like fruit pod, green drying brown	White, 1" flower with 5 purple and yellow marked petals guiding to center of flower

NOTES

Crooked spreading crown, often self-propagating, found in alleyways and byways

MAY BE CONFUSED WITH

PAULOWNIA · Velvety, larger leaves; purple, tube-like flowers with oval capsule fruits? It could be a Paulownia!

DOGWOOD GENERA · *Cornus spp.*

LEAF ARRANGEMENT	LEAF PERSISTENCE	TREE SHAPE
Simple, Opposite	Deciduous	Broad Branching

LEAF
Oval, rounded base tapers to fine point, 4 ½" leaves, veins curve to follow edge, smooth, dark green above, whitish and softly hairy below

BARK
Red-brown to blackish-gray, smooth, breaking into small square blocks (like alligator skin) upon maturity

FRUIT
Vary depending on species/cultivar

FLOWER
Showy, white or reddish-pink with 4 petals

NOTES
Relatively small trees with short trunks that branch low, lending a candelabra appearance; conspicuous flower buds at twig tips

---

FLOWERING DOGWOOD · *Cornus florida*

DISTINGUISHING FEAUTRES
4 ½" leaves, red drupe fruit in clusters of 3–5 (*see image c.*), maturing in fall and with a large, rounded, 4-petaled white or red-pink flower (*see image a.*)

KOUSA DOGWOOD · *Cornus kousa*

DISTINGUISHING FEAUTRES
4" leaves, round pink or red fruits that are upright (*see image d.*) and a pointed 4-petal white flower (*see image b.*) and camo bark (*see image e.*)

ALTERNATE-LEAF DOGWOOD · *Cornus alternifolia*

DISTINGUISHING FEAUTRES
Alternate 6" leaves and blue-black berries in flat-topped clusters on reddish stalks and flowers in loose, flat-topped clusters

MAY BE CONFUSED WITH
BLACKHAW VIBURNUM · Opposite, finely serrated, pinnately veined with red-tinted petioles or leaf edges, dark blue berry-like fruit with attractive white cluster of small flowers (like alternate-leaf Dogwood)? It could be a Blackhaw!

a.

b.

c.

d.

e.

FRINGE TREE • *Chionanthus virginicus*

LEAF ARRANGEMENT	LEAF PERSISTENCE	TREE SHAPE
Simple, Opposite	Deciduous	Broad Round, Shrublike

LEAF		BARK
6" oval leaves with smooth wavy edge, dark green with purplish petioles		Narrow and shallow ridges and furrows

FRUIT	FLOWER
Dark purple, fleshy, oval, ¾", olive-like	Fragrant, delicate, white, 4 petals reaching ¾–1" long, hanging in loose clusters

MULBERRY · *Morus spp.*

LEAF ARRANGEMENT	LEAF PERSISTENCE	TREE SHAPE
Simple, Alternate	Deciduous	Broad Round, Spreading

LEAF	BARK
Variable leaf shapes, coarse teeth, sandpapery with 1–3 lobes from ovate, mitten shape to 3-lobed (like dinosaur foot), 3–5 ½", glossy, underleaf hairy and 3 main veins from base	Pale orange-brown with small slits when young; broad, irregularly scaled with long ridges with maturity

FRUIT	FLOWER
Blackberry-like, 1–1 ¼", white, red, or black depending on species, maturing in summer	Narrow, pale green dangling catkins 1–2 ¼"

MAY BE CONFUSED WITH

OSAGE-ORANGE · Simple, shiny, tapering to a point, 2–5" leaves with a large, round 4–5" citrus-smelling, brain-like fruit with irregular, furrowed, orange-brown bark and irregular branches? It could be an Osage-Orange!

PERSIMMON · *Diospyros virginiana*

LEAF ARRANGEMENT	LEAF PERSISTENCE	TREE SHAPE
Simple, Alternate	Deciduous	Broad Irregular, Oval

LEAF	BARK
Oval, wavy edges, pointed 4 ½" leaf, smooth leaf edge, dark green above; pale, whitish below	Brown-gray, smooth turning darker, developing square plates (resembling charcoal) upon maturity

FRUIT	FLOWER
Plum-like green, orange to purple 1 ½" nearly round edible fruit	Greenish-white, ½" flower appearing in early summer

MAY BE CONFUSED WITH

BLACKGUM · Leaves appearing whorled with occasional shallow lobes, fruit is purple-blue, small, fleshy, ½"; bark with shallow furrows? It's a Blackgum! (P. 49)

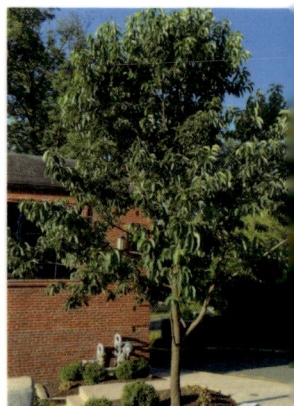

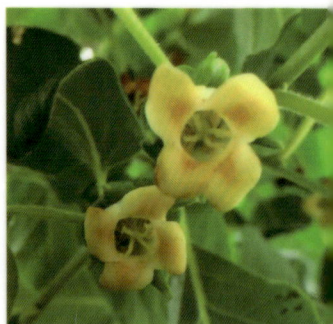

SWEETGUM · *Liquidambar styraciflua*

LEAF ARRANGEMENT	LEAF PERSISTENCE	TREE SHAPE
Simple, Alternate	Deciduous	Oval, Pyramidal

LEAF	BARK
Star-shaped, finely serrated, 5" leaves with 5–7 rounded or pointed lobes; glossy green above, lighter below	Silvery gray, smooth becoming irregularly ridged upon maturity

FRUIT	FLOWER
Distinctive, prickly, sphere "gumballs" dangling on long stems; green, maturing to brown in fall, persisting through winter	Circular, male flowers in elongated triangular gathering, female flowers individual

MAY BE CONFUSED WITH

MAPLE · 3-5 lobes with horseshoe or coat-hanger-shaped samaras? It's a Maple! (P. 78-80)

LONDON PLANETREE · Large leaves, coarsely serrated, hairy below, exfoliating orange-yellow or gray bark? It's a London Planetree! (P. 60)

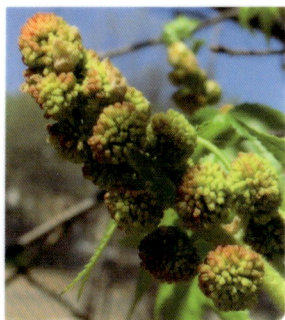

SYCAMORE · *Platanus occidentalis*

COMMON CULTIVATED VARIETY
LONDON PLANETREE · *Platanus x acerifolia*

LEAF ARRANGEMENT	LEAF PERSISTENCE	TREE SHAPE
Simple, Alternate	Deciduous	Pyramidal, Round

LEAF	BARK	FLOWER
Palmate venation, largely toothed, 5–9" leaves with 3–5 lobe , dusty green-gray above, pale and hairy below	Mottled brown bark, exfoliating into yellow-white patches. Older trees exfoliate mostly in the crown and less on the base.	Small, round, separate male and female flowers on the same tree appearing in spring before leaves

FRUIT

SYCAMORE: One circular fruit, about one inch in diameter on long stalk, covered in short hairs

LONDON PLANETREE: One, two or more circular fruits, about one inch in diameter on long stalk, covered in short hairs

NOTES

The Sycamore, native and traditionally found along riverbank environments, is one of the parents of the London Planetree that is known in urban areas because of its resistance to disease, drought, and air pollution. Look at the fruits to distinguish the two species. If you see any in pairs, it is a London Planetree!

MAY BE CONFUSED WITH

MAPLE · 3–5 lobes with horseshoe or coat hanger-shaped samaras? It's a Maple! (P. 78-80)

CRABAPPLE · *Malus spp.*

LEAF ARRANGEMENT	LEAF PERSISTENCE	TREE SHAPE
Simple, Alternate	Deciduous	Small, Broad

LEAF

Variable in shape, irregular, fine serration, pointed 2–4" leaves, pinnately veined, wavy edges, green above, underside and petiole are paler green-white usually with pubescence

BARK

Short, gray, scaly, and slightly ridged

FRUIT

Small apple-like fruit, ¼" - 2", yellow to dark red, grows on stems, fruit often persists through winter, very bitter

FLOWER

Five petals, white, light pink to dark pink, or purplish; blooms in spring

MAY BE CONFUSED WITH

HAWTHORNE · Straight thorns 1–2" and a significant number more clustered pomme fruit? It could be Hawthorne!

WEEPING WILLOW · *Salix babylonica*

LEAF ARRANGEMENT	LEAF PERSISTENCE	TREE SHAPE
Simple, Alternate	Deciduous	Round, Weeping

LEAF	BARK
Narrow, spear-shaped, serrated, 4" leaves with short stalk; yellow-green above, frosted green below	Grayish-brown, sandy, corky mature bark; rough and deep-ridged; often sprouting many pale yellow shoots from trunk

FRUIT	FLOWER
1" cluster of capsules with fine cotton-like seeds	Upright, slim, cylindrical catkin, fuzzy, 1" appearing before leaves

NOTES
Long, slender, dropping branches, curtain-like

MAY BE CONFUSED WITH

WILLOW OAK · Wider leaves, no serration along edges, acorns, and straight branches? It's a Willow Oak! (P. 43)

BEECH · *Fagus grandifolia*

LEAF ARRANGEMENT	LEAF PERSISTENCE	TREE SHAPE
Simple, Alternate	Deciduous	Oval

LEAF	BARK
Papery, pointed leaves, 4", 11–14 pairs of parallel veins ending in distinct pointed or bristle tip	Smooth, thin, gray bark often with cankers becoming defaced without ridges

FRUIT	FLOWER
Four part husk with prickles, ¾", irregularly triangular nuts (1–3 per husk) maturing in fall	Slender, 1" with fuzzy stalk, appearing just after leaves in spring

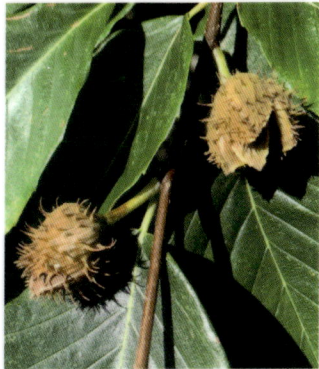

HACKBERRY · *Celtis occidentalis*

LEAF ARRANGEMENT	LEAF PERSISTENCE	TREE SHAPE
Simple, Alternate	Deciduous	Rounded crown

LEAF	BARK
Serrated, oblong leaf with sharp point, uneven leaf base, 2 ½–5" long, short leaf stalk	Young bark smooth resembling a beech, older bark develops warty, knobbed ridges

FRUIT	FLOWER
Singular, round, dark red to purple berry, ¼"	Small clusters of green male flowers and single female flowers appear at the tips of twigs in spring

NOTES
Medium to large canopy tree reaching up to 90'

RIVER BIRCH · *Betula nigra*

LEAF ARRANGEMENT	LEAF PERSISTENCE	TREE SHAPE
Simple, Alternate	Deciduous	Airy, Oval, Pyramidal

LEAF	BARK
Triangular, short-stalked, double-serrated 3" leaves with v-shaped symmetrical base; green above and whitish under leaf	Reddish, yellow, and/or gray, peeling in shaggy sheets maturing to rough, gray-orange with poorly defined ridges and blocks

FRUIT	FLOWER
Slim, hop-like flower (catkin), numerous tiny winged seeds packed between bracts, matures in spring while leaves emerge	Slender, drooping, inconspicuous flowers without petals (2–3" long)

NOTES

Most often planted as a multi-stem tree (in groups of 2–3) for aesthetic reasons, due to airy, spreading crown appearance

HORNBEAM · *Carpinus caroliniana*

LEAF ARRANGEMENT	LEAF PERSISTENCE	TREE SHAPE
Simple, Alternate	Deciduous	Oval

LEAF	BARK
Rounded base, doubly serrated, tapering to fine point, 3 ½", waxy smooth green	Smooth, bluish-gray; fluted into muscle-like ridges

FRUIT	FLOWER
Small, 1" hanging nutlet clusters carried on folded, 3-lobed, papery bracts; ripening in late summer and fall	Slim,1 ½" inconspicuous cylindrical flower catkins, pale yellow green

NOTES

Also known as ironwood, musclewood, or muscle beech

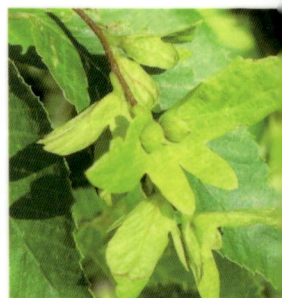

HOPHORNBEAM · *Ostrya virginiana*

LEAF ARRANGEMENT	LEAF PERSISTENCE	TREE SHAPE
Simple, Alternate	Deciduous	Oval, Round

LEAF
Rounded base, doubly serrated, tapering to fine point, 3 ½", dark green leaves, paler and fuzzy below

BARK
Smooth, reddish-brown with horizontal lenticels (slits), maturing to light brown, shredded, peeling into scales that are easily pulled off

FRUIT
Small nutlet, 1 ½–2 ½" enclosed in a hop-like sac, ripen in fall, drops with leaves

FLOWER
Slim, inconspicuous female cylindrical flower catkins and yellow-brown male catkins dangling in clusters of 3 (resembling bird toes); present throughout the winter

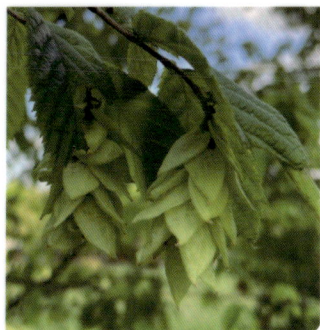

CHERRY GENERA · *Prunus spp.*

LEAF ARRANGEMENT	LEAF PERSISTENCE	TREE SHAPE
Simple, Alternate	Deciduous	Oval, Flattened Vase

LEAF	BARK
Pointed, finely serrated, 2–5" leaves, shiny, smooth, dark green above; veins on underside are fuzzy with glands present on the petiole	Gray-brown, purple tint with prominent orange-brown horizontal lenticels (resemble tiny cuts); often mangled, warty, hollowing out with age

FRUIT	FLOWER
Shiny ⅜", red, purple, or black cherry, ripe late summer–fall, large seed/fleshy fruit ratio	Large, showy, long-stalked, white to deep pink, with 5 petals; grouped in clusters along twigs; appear in early spring

JAPANESE FLOWERING CHERRIES

YOSHINO CHERRY · *Prunus x yedoensis*

DISTINGUISHING FEAUTRES
Durable single white and pale pink blossoms *(see image a.)*

KWANZAN CHERRY · *Prunus serrulata*

DISTINGUISHING FEAUTRES
Frilly double pink flowers, lack of fruit and reddish trunk *(see image b.)*

WEEPING CHERRY · *Prunus pendula*

DISTINGUISHING FEAUTRES
Drooping branches with pinkish-white blossoms *(see image c.)*

OTHER COMMON VARIETIES IN DC
Okame, Sargent, Shirofugen, Higan, Usuzumi-Zakura, Takesimensis, Akebono, Fugenzo

VIEW OUR MAP OF FLOWERING TREES IN DC: caseytrees.org/flowering

a.

c.

b.

ELM · *Ulmus americana*

LEAF ARRANGEMENT	LEAF PERSISTENCE	TREE SHAPE
Simple, Alternate	Deciduous	Oval, Flattened Vase

LEAF	BARK
Pointed, oval, doubly serrated, uneven base, 5" leaves, short, stout stalk, straight veins, dark green above, paler green to whitish below	Gray, flat, white ridges interlacing into rough, diamond-shaped fissures, peeled bark exposes whitened areas

FRUIT	FLOWER
Flattened, round, small, approximately ½", pale yellow-green samaras turn to brown; each fruit has a distinctive notch at seed bottom and tiny hairs along edges, ripen in spring	Small, greenish-red to creamy-brown, and drooping on long stalks, blooms in early spring, before leaves

NOTES

Popular tree lining streets due to its vase shape structure; popular cultivars include Jefferson, New Harmony, Princeton, Dutch, Jersey Elm, etc.

MAY BE CONFUSED WITH

HORNBEAM or **HOPHORNBEAM** · Narrow leaf with rounded base coming to a fine point; muscular or shaggy bark with brackets of seeds? It could be a Hornbeam or Hophornbeam! (P. 66-67)

LINDEN · Heart-shaped, serrated leaves with a lopsided base and winged bract? It's a Linden! (P. 71)

LINDEN GROUP · *Tilia spp.*

LEAF ARRANGEMENT	LEAF PERSISTENCE	TREE SHAPE
Simple, Alternate	Deciduous	Oval, Pyramidal

LEAF	BARK
Heart-shaped, serrated, uneven leaf base, 2 ½–7" (size dependent on species), palmate veins, green above, paler below	Gray-brown, smooth developing darker, shallow furrows and interlacing ridges

FRUIT	FLOWER
Round, nut ¼", gathered with leafy bract or wing, turning from green to yellow, 4 ½"	Pale yellow to white, ½" clustered flowers hanging with leafy bract (slender leaf) or wing

MAY BE CONFUSED WITH

ELM · Pointed, oval, 5" leaf, flat ridges, interlacing diamond, gray bark, vase-shaped tree form? It's an Elm! (P. 70)

MULBERRY · Variable leaf shape teasing lobes with pale orange-brown bark? It's a Mulberry! (P. 57)

AMERICAN BASSWOOD ·
Tilia americana

Largest, 7 ½" leaves, uneven with hairless leaf stalks

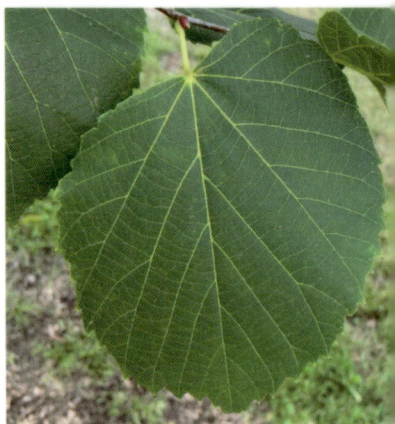

LITTLELEAF LINDEN ·
Tilia cordata

DISTINGUISHING FEAUTRES
Smaller, 2 ½" leaves with pale blue-green underleaf with orange hairs

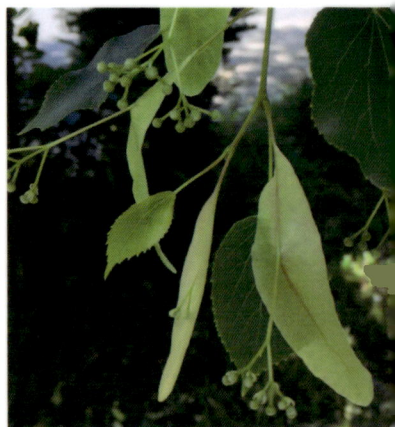

SILVER LINDEN ·
Tilia tomentosa

DISTINGUISHING FEAUTRES
Medium, 3 ½" rounded leaves, darker green above, silvery underleaf

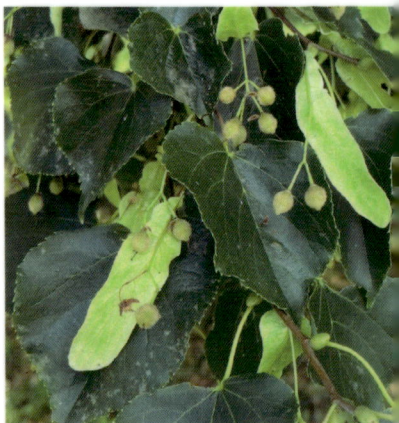

SERVICEBERRY · *Amelanchier spp.*

LEAF ARRANGEMENT	LEAF PERSISTENCE	TREE SHAPE
Simple, Alternate	Deciduous	Oval, Upright Vase

LEAF	BARK
Round, pointed, finely serrated, smooth 1 ½–3" leaf	Smooth, ashy gray with stripes, developing roughness and longer slits

FRUIT	FLOWER
Round, drooping fruit, green, red to blue-black in color when ripe, edible, appearing in early or mid summer	Clusters of upright, white, slender flowers

NOTES

Multi-stemmed shrub or small tree with narrow crown, also known as Juneberry, Shadbush, Saskatoon or Chuckley Pear

MAY BE CONFUSED WITH

SMOKETREE · Silky hair flower stalks lending "smoky" effect and rounded oval, 5" blueish-green leaves? It could be an American Smoketree!

CALLERY PEAR · *Pyrus calleryana*

LEAF ARRANGEMENT	LEAF PERSISTENCE	TREE SHAPE
Simple, Alternate	Deciduous	Oval, Columnar

LEAF	BARK
Round to heart-shaped, subtle, fine serration 2–3" leaves, shiny green above and duller below	Smooth dark gray to light brown with lenticels becoming grayish brown with neat, narrow, shallow furrows and scales

FRUIT	FLOWER
Round, ½", brown and bitter	Clusters of showy white flowers, ½", appears before leaves, masking tree in white

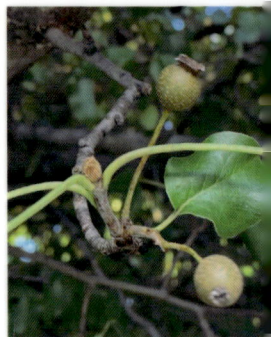

JAPANESE ZELKOVA · *Zelkova serrata*

LEAF ARRANGEMENT	LEAF PERSISTENCE	TREE SHAPE
Simple, Alternate	Deciduous	Vase, Upward Branching

LEAF	BARK
Narrow, tapering to point with rounded serration and uneven base, 4" leaves	Smooth gray to dark copper with exfoliated, cracking bark revealing patches of orange lenticels becoming mottled in appearance upon maturity

FRUIT	FLOWER
Small, triangular, fleshy fruit with thin skin, appearance of a misshapen green pea; found at base of leaves, green to brown, appearing mid to late summer	Short strings of yellowish flowers that gather along new twig growth, blooms in early spring

NOTES

Species traditionally used as a Bonsai because of hardiness and adaptability, explaining also its success as an urban tree

MAY BE CONFUSED WITH

CHINESE ELM · Smaller leaves (1 ½"), thin, broadly lanceolate, with non-rounded teeth; bark is mottled and camouflage-like, fruit is flattened, brown samara? It's a Chinese Elm! (P. 76)

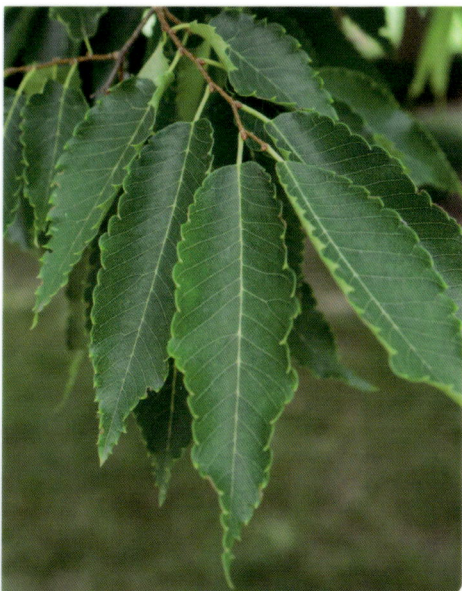

CHINESE ELM · *Ulmus parvifolia*

LEAF ARRANGEMENT	LEAF PERSISTENCE	TREE SHAPE
Simple, Alternate	Deciduous	Vase, Round

LEAF
Narrow, tapering to point with blunt serration and uneven base, 1 ½" leaves, shiny, leathery, dark green above and paler below

BARK
Mottled, camouflage-like pattern of green, gray, and orange, scales peel to reveal orange inner bark

FRUIT
Flattened, circular, ½" samara, hairless, pale yellow to light reddish-brown, with a deep notch, appears in tight clusters in the fall

FLOWER
Inconspicuous, small, tight clusters of light red-green flowers by leaf axils, appear in fall

MAY BE CONFUSED WITH

JAPANESE ZELKOVA · Larger leaves (4" long), with rounded teeth; bark is mottled with horizontal lenticels (small slits); fruit is small, triangular, and looks like a misshapen green pea? It's a Japanese Zelkova! (P. 75)

SIBERIAN ELM · Leaves similar size (1 ½"), sharper serration, larger, rounded buds with unmottled, dark, furrowed bark? It could be a Siberian Elm!

YELLOWWOOD · *Cladrastis kentukea*

LEAF ARRANGEMENT	LEAF PERSISTENCE	TREE SHAPE
Compound, Odd-Pinnate, Alternate	Deciduous	Broadly Round

LEAF	BARK
8–12" long, with 5–11 lemon/eye shaped 3" leaflets, distinct venation, green above, paler below	Smooth, thin, sometimes wrinkled, pale gray, often with lichens and mosses

FRUIT	FLOWER
Flat pod, 3", green ripening to brown in fall, sometimes persisting into winter	Drooping in terminal hanging clusters, pea-like flower, creamy/white in color, 10", mildly fragrant and appearing in late spring to early summer

MAY BE CONFUSED WITH

DOGWOODS · Large leaf, veins curving to follow leaf edges; bark breaking into small square blocks; and red fruit or blue-black berries? It's a Dogwood! (P. 54)

BLACKGUM · 4" long, oval with blunt tip, appearing whorled around the branch, some leaves turning red; gray-brown bark; and blue-black berries? It's a Blackgum! (P. 49)

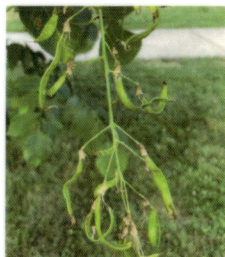

RED MAPLE · *Acer rubrum*

LEAF ARRANGEMENT	LEAF PERSISTENCE	TREE SHAPE
Simple, Opposite	Deciduous	Oval, Upright

LEAF

Relatively small, serrated 4" leaves, 3 shallow lobes with long red leaf stalk, green above, whitened and sometimes hairy below, and turning red or orange in autumn

BARK

Smooth, dry, sandy ridges and light gray becoming darker, breaking into narrow, scaly plates, forming ridges upon maturity

FRUIT

Samara with v-shaped wings spreading at a narrow angle, less than 1", reddish-brown often with brighter red edges

FLOWER

Attractive, small, red, hanging in clusters appearing before leaves in spring

SUGAR MAPLE · *Acer saccharum*

LEAF ARRANGEMENT	LEAF PERSISTENCE	TREE SHAPE
Simple, Opposite	Deciduous	Oval, Round

LEAF	BARK
Symmetrical from center lobe, few teeth, 5" leaves with 5 U-shaped lobes, short basal lobe	Gray-brown, smooth, developing sandy furrows with irregular plates

FRUIT	FLOWER
Horseshoe-shaped samara, papery wings are almost parallel; wings turn brown while seeds remain green as they ripen in fall	Pale yellow-green, 1–3", slender stems; appear with or before leaves in early spring

SILVER MAPLE · *Acer saccharinum*

LEAF ARRANGEMENT	LEAF PERSISTENCE	TREE SHAPE
Simple, Opposite	Deciduous	Broad, Oval

LEAF	BARK	FRUIT
Deep lobes/sinuses with lacy-like serrations, 6 ½", 5 lobes, silvery underleaf	Silvery, turning shaggy with narrow strips	Wide spreading, samara 1 ¾"

NOTES

Feathery branches, deeply cut leaves, Red x Silver Maple hybrids cultivated frequently such as "Freeman" Maple

BLACK LOCUST · *Robinia pseudoacacia*

LEAF ARRANGEMENT
Compound, Odd-Pinnate, Alternate

LEAF PERSISTENCE
Deciduous

TREE SHAPE
Broadly Round

LEAF
11" long, with 7–19 oval/grape-like 1" leaflets, green above, paler gray-green below

BARK
Gray to light brown, scaly ridges developing thick, corky look with deep furrows and interlacing ridges

FRUIT
Flattened pods, 3", smooth, green to brown, enclosing red-brown kidney-like beans, persisting through winter

FLOWER
Showy, white, fragrant, pea-like flower hanging in 5" clusters, appearing in late spring

NOTES
Branches have zig-zag appearance, some twigs with thorns.

JAPANESE PAGODA · *Sophora japonica*

LEAF ARRANGEMENT	LEAF PERSISTENCE	TREE SHAPE
Compound, Odd-Pinnate, Alternate	Deciduous	Oval

LEAF	BARK
8" long, with 7–17 oval, pointed 1–2" leaflets, green above, paler below	Brown-gray with reddish-brown long furrows and vertical ridges

FRUIT	FLOWER
Neon green-yellow legume turning brown, 5", hanging in clusters like a string of pearls, maturing in early fall but persisting through winter	Creamy white, pea-like flower in clusters in summer

CHINESE PISTACHE · *Pistacia chinensis*

LEAF ARRANGEMENT	LEAF PERSISTENCE	TREE SHAPE
Compound, Odd-Pinnate, Alternate	Deciduous	Rounded Oval

LEAF	BARK
10" long, with 8–16 narrow, tapering to point 2–4" leaflets	Brown and gray bark with reddish shallow furrows developing rectangular flat square ridges

FRUIT	FLOWER
Red or blue round fruit, ½", grape-like in stem clusters, appear in early fall, and fruit persists at twig ends through winter	Ashy red to green, 2–3", clustered, somewhat showy

MAY BE CONFUSED WITH

TREE OF HEAVEN · Twisted, papery samaras? It's a Tree of Heaven!

BLACK WALNUT • *Juglans nigra*

LEAF ARRANGEMENT	LEAF PERSISTENCE	TREE SHAPE
Compound, Pinnate, Alternate	Deciduous	Round

LEAF	BARK
1–2' long, with 10–24 narrow, pointed, finely serrated 4" leaflets, small or absent terminal leaflet, green to yellow-green, paler below	Brown turning darker with diamond furrows

FRUIT	FLOWER
Almost round, 2" thick green husk fruit, hard nut with irregular furrows and edible, oily, meaty seeds	Male flower as dangling catkins, 4", emerging with leaves while female flower small, green tip developing with shoots

MAY BE CONFUSED WITH

CHINESE PISTACHE • Smaller leaves 2–4", fruit small, red or blue, ½" on clustered stems? It's a Chinese Pistache! (P. 83)

KENTUCKY COFFEETREE • Larger, wider leaves (1 ½–2") with thick, flat seed pod? It's a Kentucky Coffeetree! (P. 91)

BOXELDER · *Acer negundo*

LEAF ARRANGEMENT	LEAF PERSISTENCE	TREE SHAPE
Compound, Odd-Pinnate, Opposite	Deciduous	Irregular Round

LEAF	BARK
6–8" long, with 3–5 (sometimes 7) pointed oval/ serrated, sometimes lobed 2–4" leaflets; light green, paler below	Gray to light brown, lightly ridges when young, turning furrowed with shallow ridges

FRUIT	FLOWER
Drooping clustered of paired V-shaped samaras, 1 ¾", turning tan and persisting through winter	Long, drooping flowers, pale yellow-green or red in spring

NOTES

Only Maple with compound leaves and common in forested areas.

MAY BE CONFUSED WITH

POISON IVY · 3 leaves, irregularly toothed, red connecting stem, forest carpet or climbing vine present on tree? Watch out and don't touch! It could be Poison Ivy!

ASH · Pointed oval, usually 7 finely serrated leaflets with ashy corky bark and one-winged, flat samara fruit? It's an Ash! (P. 86)

ASH GENERA · *Fraxinus spp.*

LEAF ARRANGEMENT	LEAF PERSISTENCE	TREE SHAPE
Compound, Pinnate, Opposite	Deciduous	Oval

LEAF	BARK
Pointed oval, smooth, or finely serrated 4" leaflets, with usually 7 (but anywhere from 5–13) leaflets (6–9"), hairless, green, paler below	Ashy gray-brown, corky ridges interlacing, forming diamond pattern

FRUIT	FLOWER
One-winged, 1 ½" long, flattened samara with formed seed cavities, ripe fruit green drying to light or dark brown in fall	Male flowers in green-purple clusters, female flowers in green spray

NOTES
Tree species threatened by Emerald Ash Borer (beetle)

MAY BE CONFUSED WITH
BLACK WALNUT · 15–19 narrow, thin leaflets, shiny green, finely serrated; bright green, smooth, circular nut (1–2"); gray-brown, flat, scaly ridges? It's a Black Walnut! (P. 84)
AMUR CORK TREE · 7–11 leaflets, broader, dark green, pale below; thick, corky, furrowed ridges; black, grape-like fruit clusters; flat-topped crown? It could be an Amur Cork Tree!

BLACK ASH ·
Fraxinus nigra

DISTINGUISHING FEAUTRES
Leaflets without stalks

CAROLINA ASH ·
Fraxinus caroliniana

DISTINGUISHING FEAUTRES
Medium, 3 ½" rounded leaflets, darker green above, silvery underleaf

GREEN ASH ·
Fraxinus pennsylvanica

DISTINGUISHING FEAUTRES
Leaflet that narrows gradually at base, underleaf not as white as White Ash

WHITE ASH ·
Fraxinus americana

DISTINGUISHING FEAUTRES
U-shaped leaf scars and whitened underside

BUCKEYE · *Aesculus glabra*

LEAF ARRANGEMENT		LEAF PERSISTENCE	TREE SHAPE
Compound, Palmate, Opposite		Deciduous	Oval

LEAF	BARK
Large leaves, 9", 5–7 leaflets, serrated, more elongated and coming to fine point, dark green above and paler below	Light brown-gray, getting darker, with corky, scaly, and gray with large ridges

FRUIT	FLOWER
Dependent on species but generally thick, smooth, or spiny husk with shiny, brown nut, highly toxic if ingested	Yellow in erect clusters, 6" long, appear in spring

NOTES

Stout, smaller, orange-brown buds

MAY BE CONFUSED WITH

HORSE CHESTNUT · Fatter, blunt leaves with similar fruit? It could be a Horse Chestnut!

HONEY LOCUST · *Gleditsia triacanthos*

LEAF ARRANGEMENT		LEAF PERSISTENCE	TREE SHAPE
Compound, Pinnate, Alternate		Deciduous	Oval, Vase

LEAF	BARK
7" long, with 15–30 small/elliptical ½–1 ½" leaflets, pinnately or bipinnately compound, dark green to yellow-green	Brown-gray with orange tint, lenticels present developing broad, peeling, and curling ridges upon maturity, thorns often present on trunk

FRUIT	FLOWER
Flattened, curling seed pod, 6–8", turning brown, with dark, shiny seed, ⅓"	Small, inconspicuous green-yellow flowers on short stalks, hanging in clusters, in late spring, fragrant

NOTES
Zig-zag twigs with small knobs.

MAY BE CONFUSED WITH

MIMOSA · Tiny ⅜" leaflets, feathery, air canopy appearance, stingy pink flowers, green stem, self propagating? It's a Mimosa!

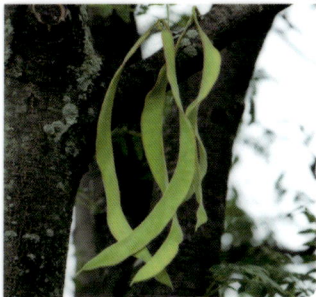

GOLDEN RAINTREE · *Koelreuteria paniculata*

LEAF ARRANGEMENT	LEAF PERSISTENCE	TREE SHAPE
Compound, Odd-Pinnate, Alternate	Deciduous	Round, Vase

LEAF	BARK
12" long, with 7–17 irregularly serrated and lobed leaflets, deep green above, lighter below	Silver-gray, reddish-brown, shallow

FRUIT	FLOWER
Papery, 3–sided, triangular capsule, 1 ½", hanging in clusters	Ornamental, bright yellow, 10–15" large spray of flowers

MAY BE CONFUSED WITH

CHINABERRY · Single to double compound, 1–2" leaves, leaflets serrated or lobed (10–22") with yellow-brown, round, ¾" fruit in clusters. It could be a Chinaberry!

KENTUCKY COFFEETREE • *Gymnocladus dioicus*

LEAF ARRANGEMENT	LEAF PERSISTENCE	TREE SHAPE
Compound, Bi-Pinnate, Alternate	Deciduous	Upright

LEAF	BARK
1–3' long, with 6–15 large/oval pointed 1 ½–2" leaflets and 5–9 pairs of leaflets	When young, pale gray with shallow ridges and orange furrows becoming deeper and scaly upon maturity

FRUIT	FLOWER
Orange-brown, thick, 5", flat seed pod	Small, white, clustered flowers at twig tips

MAY BE CONFUSED WITH

BLACK WALNUT · 15–19 narrow, thin leaflets, shiny green, finely serrated; bright green, smooth, circular nut (1–2"); gray-brown, flat, scaly ridges? It's a Black Walnut! (P. 84)

TREE OF HEAVEN · Narrower leaves, 11–41 leaflets, twisted, papery samaras? It's a Tree of Heaven!

SPECIES INDEX